GEOLOGICAL FOSSIL ENCE

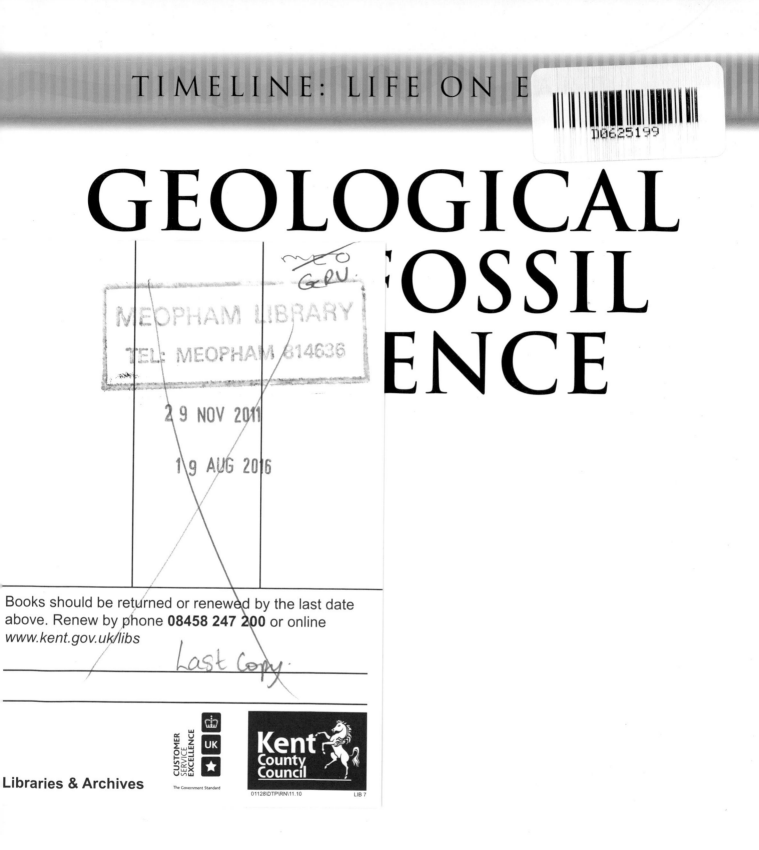
MICHAEL BRIGHT

Heinemann
LIBRARY

www.heinemannlibrary.co.uk
Visit our website to find out more information about Heinemann Library books.

To order:
☎ Phone +44 (0) 1865 888066
🖷 Fax +44 (0) 1865 314091
🖳 Visit www.heinemannlibrary.co.uk

Heinemann Library is an imprint of Capstone Global Library Limited, a company incorporated in England and Wales having its registered office at 7 Pilgrim Street, London, EC4V 6LB – Registered company number: 6695582

"Heinemann" is a registered trademark of Pearson Education Limited, under licence to Capstone Global Library Limited.

Text © Capstone Global Library Limited 2009
First published in hardback in 2009
Paperback edition first published in 2010
The moral rights of the proprietor have been asserted.

Edited by Pollyanna Poulter
Designed by Steven Mead and Q2A Creative Solutions
Original illustrations © Pearson Education Limited by International Mapping and Jeff Edwards
Picture research by Elizabeth Alexander
Production by Alison Parsons
Originated by Dot Gradations
Printed in China by Leo Paper Group

ISBN 978 0 431064 73 4 (hardback)
13 12 11 10 09
10 9 8 7 6 5 4 3 2 1

ISBN 978 0 431064 79 6 (paperback)
14 13 12 11 10
10 9 8 7 6 5 4 3 2 1

British Library Cataloguing-in-Publication Data
Bright, Michael
    Geological and fossil evidence. - (Timeline : life on Earth)
    1. Fossils - Juvenile literature 2. Paleontology - Juvenile literature 3. Historical geology - Juvenile literature
    I. Title
    560
A full catalogue record for this book is available from the British Library.

Acknowledgments
We would like to thank the following for permission to reproduce photographs: © Alamy: pp. **10** (Russ Bishop), **32** (blickwinkel), **44–45** (Natural Visions); © Ardea: p. **8** (Francois Gohier); © Corbis: pp. **5** (Maurice Nimmo/ Frank Lane Picture Agency), **14** (Roger Ressmeyer), **20** (Naturfoto Honal), **28** (Anthony Bannister/Gallo Images), **34** (Frans Lanting), **35** (Layne Kennedy), **41** (Louie Psihoyos), **45** (Jeffrey L. Rotman), **46** (Reuters); © Getty Images: pp. **9**, **39**, and **43** (Louie Psihoyos/Science Faction), **19** (Spencer Platt), **26** (Michael Melford/National Geographic), **33** (Dorling Kindersley), **42** (Pete Oxford/ Minden Pictures), **47** (AFP); © imagequestmarine.com: p. **37** (Johnny Jensen); © Istockphoto: pp. **4**, **14**, **24**, and **36 chapter openers**, and all **panel backgrounds** (Heiko Grossmann); © NHMPL: p. **21** (De Agostini); © Photolibrary: p. **29** (OSF/Harold Taylor); © Robert Nicholls: p. **31**; © Science Photo Library: pp. **6**, **11** (Chris Butler), **16** (Susumu Nishinaga), **17** (Christian Darkin), **18** (Friedrich Saurer), **22** (Pascal Goetgheluck), **23** (Philippe Plailly/ Eurelios), **25** (Georgette Douwma), **30** (Sinclair Stammers), **40** (Joe Tucciarone); © Simon Conway Morris, University of Cambridge: p. **27**.

Cover photograph of cast of Archaeopteryx fossil remains reproduced with permission of © Jim Amos (Science Photo Library), Earth from space © NASA.

We would like to thank Prof. Norman MacLeod for his invaluable help in the preparation of this book.

Every effort has been made to contact copyright holders of material reproduced in this book. Any omissions will be rectified in subsequent printings if notice is given to the Publishers.

Disclaimer
All the Internet addresses (URLs) given in this book were valid at the time of going to press. However, due to the dynamic nature of the Internet, some addresses may have changed, or sites may have changed or ceased to exist since publication. While the author and Publishers regret any inconvenience this may cause readers, no responsibility for any such changes can be accepted by either the author or the Publishers.

CONTENTS

Some words are printed in bold, **like this**. You can find out what they mean in the glossary.

FOSSILS

How can we find out about the age of Earth and the **organisms** that lived on it before us? We use evidence found in rocks, including fossils. Fossils are the preserved remains or traces of past life. They can be formed from a part of an organism, such as bones, shells and leaves, or consist of the entire organism preserved in rocks or **sediments**. This is called a "**body fossil**". The preserved remains can also be a footprint, track, or burrow, known as a "**trace fossil**". Unfortunately, the many factors affecting preservation mean that most organisms die without leaving any trace at all.

Few fossils

Fossilization is rare and only occurs when conditions are just right. We may never know anything about many of the organisms that lived in the past simply because they never became fossils. An animal that lived in the ocean is more likely to be fossilized than one that lived on land. Similarly, an animal that died in a river is more likely to be fossilized than one that died on dry land. This is because sediments are laid down more frequently in the sea and in rivers. As a first step towards fossilization the body must be covered in sediment soon after death in order to prevent decay. When buried it

16TH CENTURY
Leonardo da Vinci observes that the Biblical flood failed to explain the origins of fossils.

1676
Dinosaur thighbone found by the Reverend Plot in England.

18TH CENTURY
William Smith becomes the "father" of English geology.

1778
Swiss geologist Jean-André Deluc first uses the term "geology".

| 1500 | 1600 | 1700 |

must not be disrupted or destroyed by physical, chemical, or biological processes. A creature with hard parts, such as an oyster, is much more likely to be fossilized than a soft-bodied one, such as a jellyfish. Due to the complicated processes involved in fossilization the evidence of life that we have found in rocks, called the **fossil record**, is sparse.

Zombie effect

Palaeontologists can be tricked, however. A hard fossil, such as a dinosaur tooth, might be washed out of one layer of **sedimentary rocks** and then re-fossilized in newer sediments laid down millions of years later. It would be a mistake to believe that the owner of the tooth lived at the same time as the younger rocks. This is called the "**zombie effect**" or "reworking", when fossils become "walking dead". They move from a rock layer of one age and end up in a rock layer of a different age.

The fossilized leaves of an extinct seed fern. This is in fact not a true fern, but is an early gymnosperm tree. It reproduced using seeds, rather than spores as ferns do.

FOSSIL RECORD

Geologists know that **sedimentary rocks** are laid down one on top of the other, so the fossils in them will generally be in date order, with the oldest at the bottom and the youngest at the top. This rule applies as long as the rocks have not been turned upside down in a geological upheaval. It means that palaeontologists (scientists who study fossils) have a record showing the **succession** of organisms – the order in which organisms were alive. By looking at the fossils they can see how and where an older type of organism evolved into a younger type of organism, and when this occurred. It is evidence like this that supports the theory of evolution.

1809
Encyclopaedia Britannica includes the term "geology".

1824
English academic William Buckland discovers and names *Megalosaurus*.

1842
The term "dinosaur" is coined by British anatomist Richard Owen.

1858
William Parker Foulke discovers the first almost complete dinosaur skeleton in New Jersey, USA.

1800   1850   1900

NICOLAS STENO: FOSSIL PIONEER

The naturalists of ancient Greece and Rome found what they thought were magical objects embedded in rock. They called them "tongue stones". Roman author and natural philosopher Pliny the Elder (AD 23–79) suggested that the stones had fallen from the sky. Legend had it that they were the tongues of serpents, and this was the general belief until October 1666, when Danish clergyman and anatomist Nicolas Steno (1638–1686) was presented with the head of a great white shark. When he dissected it, he noticed that its teeth closely resembled tongue stones. He deduced that the stones were in fact the teeth of ancient sharks – they were fossils.

In Steno's time, the word "fossil" was applied to anything that came from the earth, including crystals. Steno showed that some of these "fossils" were once living organisms. It led him to the question of how a solid object could become embedded in solid rock. He went on to describe how he thought the layers of sedimentary rocks were laid down, with younger ones on top of older ones. Steno also noticed that rocks in some layers might be rich in fossils while others were barren. It was one of the first times geology and fossils had been used to identify different time periods in Earth's history.

▲ *Nicolas Steno was not alone in realising fossils were once living things. English scientist Robert Hooke (1635–1703) and English natural historian John Ray (1628–1705) were reaching the same conclusion.*

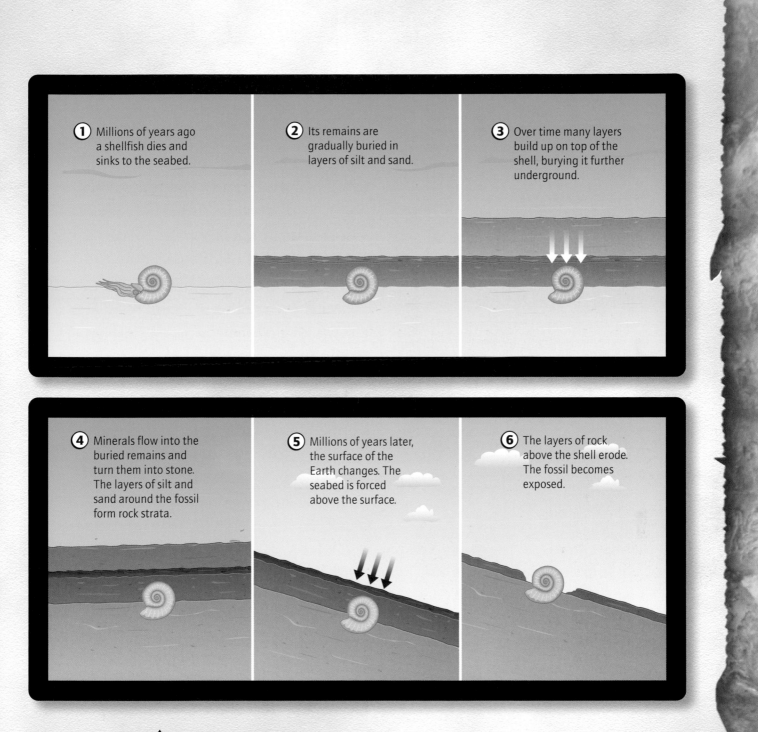

1 Millions of years ago a shellfish dies and sinks to the seabed.

2 Its remains are gradually buried in layers of silt and sand.

3 Over time many layers build up on top of the shell, burying it further underground.

4 Minerals flow into the buried remains and turn them into stone. The layers of silt and sand around the fossil form rock strata.

5 Millions of years later, the surface of the Earth changes. The seabed is forced above the surface.

6 The layers of rock above the shell erode. The fossil becomes exposed.

▲ *This series of pictures shows the process of fossilization, during which a living organism turns into a fossil trapped in rocks.*

THE WORLD'S BEST PRESERVED DINOSAUR

▲ *Many hadrosaurs had peculiar hollow, bony crests on their heads. It is thought they were used to produce foghorn-like calls.*

There are times in science when luck plays a great part in new discoveries. Luck was certainly on the side of 16-year-old high school student Tyler Lyson. In 1999, he stumbled upon three vertebrae from a hadrosaur, or duck-billed dinosaur. They were protruding from a layer of rock that was laid down in an ancient river floodplain in North Dakota, USA. He noted its position and thought no more about it; after all, duck-billed dinosaur fossils are relatively common. They were the "cattle" of the **Cretaceous**, large herbivores that wandered about in herds around 67 million years ago.

Chance, however, took Tyler back to the site. Upon finding a piece of fossilised skin Lyson realised he had something special.

Dinosaur surprise

Dr. Phillip Manning of Manchester University, UK, oversaw the hadrosaur's excavation. The entire skeleton of an 8 m (25 ft.) long *Edmontosaurus* was taken out in two enormous blocks of rock. They were examined in Los Angeles, USA, using a **CT scanner** that NASA and Boeing normally use to test space shuttle parts. It enabled scientists to see the dinosaur inside the rock. Soft tissue, such as muscles, tendons, and skin, was visible under the scanner, giving new information about body size and the way in which the dinosaur worked.

The hadrosaur's skin had not shrunk against its bones so scientists could see the animal's true proportions. One surprise was that its rump was about 25 percent bigger than expected. Calculations based on its potential muscle power indicate that it could run at roughly 45 kmh (28 mph), and therefore outrun many predators, such as *Tyrannosaurus rex*. The hadrosaur had two feet with hooves made of keratin, a material similar to that which your fingernails are made from. This indicates that the dinosaur probably walked on two legs, not all four. Two digits on each hand were joined together with skin, suggesting that they might have been used for swimming. The vertebrae were set about 1 cm (0.4 in.) apart, which means that scientists need to reassess the length of dinosaurs. In museums, skeletons are usually seen with vertebrae joined tightly together. This revelation could add up to 1 m (3.3 ft.) to the length of the larger dinosaurs. The skin of the hadrosaur shows signs of stripes, similar to those of a zebra. These stripes could have been used for camouflage or to serve a social function, such as indicating an animal's identity by its pattern of stripes.

Scavengers

Lyson's hadrosaur is thought to have fallen into a river in which it was covered rapidly with silt. Another surprise was that a scavenging crocodile probably began to feed on the dinosaur, but somehow got caught up in the carcass and died too. Its forelimb is poking through the dinosaur's chest – a fossil within a fossil.

▲ *Building a dinosaur from fossilized bones that have been dug out of the ground is like completing a gigantic, three-dimensional jigsaw puzzle. Each bone (above) is wrapped in a protective plaster coat.*

Geological time

As we have seen, in the 17th century, Nicolas Steno recognized rock succession – that younger rocks are laid down horizontally upon older rocks – but it was not until the 19th century that English surveyor and canal builder William Smith took this idea further. He noticed that when he dug a canal, the fossils in different rock layers were always in the same order from bottom to top, and that the same succession of fossils could be found elsewhere. He realised that the fossils contained in those rocks could be used not only to match the different locations in which they occurred, but also to date them. This layering of rocks, known as **stratigraphy**, is the principle behind one method of dating that is used to this day.

Divisions of the geological timescale

Once geologists had recognized that different layers of fossils represented different ages, they could draw up a table to show the order. They called it a stratigraphic column. It was divided into time zones that contain similar fossils. Some zones were named after the places where rocks of a particular age were first studied. The **Permian** is named after the province of Perm in Russia and the **Cambrian**

A hiker stands next to twisting layers of sandstone known as "The Wave", located on the Colorado Plateau near the Arizona-Utah border, USA. The blends of colour are caused partly by deposits of iron.

DATING ROCKS AND FOSSILS

Today, several methods are used to date rocks. **Elements** found within rocks change over time and measuring those changes provides geologists with a "clock" that can be used to date them. These are two of the most important methods:

- Radiometric carbon dating can be used to date anything that was once living (organic) and less than 70,000 years old.

- **Radioisotopic dating** compares the relative abundance of two **isotopes** in rocks, such as **potassium** and **argon**, **uranium** and lead, and **rubidium** and **strontium**. This can be used to date rocks of any age.

By using different pairs of isotopes a comparison can be made to check the accuracy of the date. The dating of younger rocks tends to be more accurate than older rocks.

after the Roman name for Wales, where rocks of this date were first found. Others were named after special features of the rocks. For instance, the Cretaceous period is named after the Latin word for "chalk".

Geologists also noticed that the fossilized organisms became more complex as time went on. They saw that the very oldest rocks had no fossils, slightly younger rocks had simple marine creatures, and recent rocks contained more complex animals such as fish, amphibians, reptiles, birds, and eventually mammals. At the boundaries between the major geological periods were times when many fossils disappeared suddenly from the fossil record. These "mass extinctions" killed off whole groups of living things. Geologists noted that each mass extinction was followed by periods of rapid "**adaptive radiation**", when the survivors of the extinction event evolved into many new forms and filled all the vacant **ecological niches**.

There is a change towards greater complexity in the fossil record that became clearer in 1859 when Charles Darwin published *On the Origin of Species*. It showed that succession in the fossil record could be used to demonstrate the evolution of life over the course of 4 billion years.

Very often, rocks are not found in the same part of the globe where they were laid down. They move around the surface of Earth on the drifting continents.

A geological timeline

The great journey of life from its humble beginnings to the complexity of humans can be divided up into recognizable time periods. Particular organisms dominated many periods, often called the "great ages". The age of the dinosaurs, for example, lasted for roughly 165 million years, and occurred long before humans arrived. By contrast, humans have only been around for a mere 200,000 years, yet it has been suggested that an **epoch** be named after us – the "Anthropocene" – to begin in 1800 BC. This is because in the short time that humans have been on Earth we have influenced events greatly, especially since the start of the **Industrial Revolution**.

THE GEOLOGICAL TIMELINE

Eons and Times	Eras	Periods	Epochs	Years	Events
Phanerozoic	Cenozoic	Quaternary	Holocene	today–11,430 years ago	Holocene extinction. Age of humans.
			Pleistocene	11,430 years ago–2.59 MYA	Sabre-toothed cats and mammoths.
		Tertiary	Pliocene	2.59–5.33 MYA	Hominids (ancestors of humans), many whales, and *Megalodon* shark.
			Miocene	5.33–23.03 MYA	Horses, dogs, bears, modern birds.
			Oligocene	23.03–33.9 MYA	Rhinos, deer, pigs.
			Eocene	33.9–55.8 MYA	Rodents, whales.
			Palaeocene	55.8–65.95 MYA	Large mammals, primitive primates.
	Mesozoic	Cretaceous		65.95–145.5 MYA	Cretaceous-Tertiary extinctions. Marsupials, snakes, crocodiles, bees, butterflies, flowering plants.
		Jurassic		145.5–199.6 MYA	Birds, pterosaurs. Age of dinosaurs.
		Triassic		199.6–251 MYA	Triassic-Jurassic extinctions. Mammals, ichthyosaurs, dinosaurs.

Eons and Times	Eras	Periods	Epochs	Years	Events
	Palaeozoic	Permian		251–299 MYA	Permian mass extinctions. Age of amphibians.
		Carboniferous		299–359.2 MYA	Reptiles, winged insects, conifers. Golden age of sharks
		Devonian		359.2–416 MYA	Late-Devonian extinctions. Amphibians, sharks, club mosses, horsetails, ferns. Age of fishes.
		Silurian		416–443.7 MYA	Vascular plants, jawed fish.
		Ordovician		443.7–488.3 MYA	Ordovician-Silurian extinctions. Land plants, corals.
		Cambrian		488.3–542 MYA	Jawless fishes, vertebrates. Age of **trilobites**. **Cambrian explosion** – today's major groups of animals or **phyla** appear.
Precambrian time	Proterozoic			542–2,500 MYA	Sponges. Multicellular life.
	Archaean			2,500–3,800 MYA	Blue-green algae, bacteria. Life forms in sea.
	Hadean			3,800–c.4,570 MYA	Earth's crust solidifies.

Earth, together with the rest of the solar system, was formed about 4.56–4.57 billion (456–457 thousand million) years ago.

Shaping Earth

Did you know that fossils of sea creatures are found at the rocky top of Mount Everest, precisely 8,844.43 m (29,017 ft.) above sea level? How could this have happened? In order to understand the history of Earth, we must first understand the processes that have shaped it.

The surface of Earth is never still. There are major geological forces continually at work. All the continental landmasses, or plates, "float" on the upper **mantle** and continually shift positions. This slow process is known as **continental drift** or **plate tectonics**. One plate can be forced down below another,

giving rise to explosive volcanoes, such as those in Indonesia, including the infamous Krakatoa which erupted in 1883. Alternatively, one plate can slam into another, buckling Earth's crust to push up great mountain chains, such as the Himalayas and Alps. Of course, what goes up must eventually come down. Ice, water, and wind all erode the mountains, with the debris from them settling on the bottom of the sea, in rivers or on the desert floor to form new rocks. These rocks can be pushed right up again by more geological upheavals to form new mountains. This is why fossilized sea creatures are found at the top of Mount Everest.

Fossils and rocks

There are three main types of rock: igneous, sedimentary, and metamorphic.

Igneous rocks first appear as molten rock, known as **magma**, that wells up from Earth's interior. It can form large reservoirs below ground that cool to form rocks such as **granite**. Magma can also burst through the surface from volcanoes, when it is known as lava. It cools to form rocks such as **basalt**. Igneous rocks do not contain fossils, but they can be used to date other rocks above and below them. Layers of volcanic ash, for example, settle and solidify as **tuffs**. Beneath them there might be layers of **sedimentary** rock that are rich in fossils, and more sedimentary rocks might have formed on top of the tuffs. The tuff layer is therefore sandwiched by sedimentary rocks and can be used to date the layers that appear directly

In California, the movements of the continents can be seen first-hand along the San Andreas Fault, the boundary between the Pacific Plate and the North American Plate. Rocks on the west side of the fault are moving slowly to the northwest.

above and below it. This method has been used to date sediments containing early human fossils in East Africa.

Sedimentary rocks

Sedimentary rocks are formed when layers of sediment are deposited one on top of the other. The bottom layers are compressed over a period of time and slowly turn into rock. Most fossils are found in sedimentary rocks, where they accumulate along with the sediments. Sandstones, for example, are made from tiny particles of other rocks that settled on the bottom of rivers, lakes, or seas or in the desert. Mummified dinosaurs have been found in sandstone rocks. In some cases, sedimentary rocks consist mainly of solidified animal remains. Certain types of limestone are made of compressed animal bodies, such as corals.

Metamorphic rocks

Metamorphic rocks have changed from one kind of rock into another. They were once igneous or sedimentary rocks, but they have been changed by intense heat and pressure. Fossils can be found in metamorphic rocks, as long as they have not been altered too much, but they are rare because the heat and pressure that forms metamorphic rocks generally destroys the fossils.

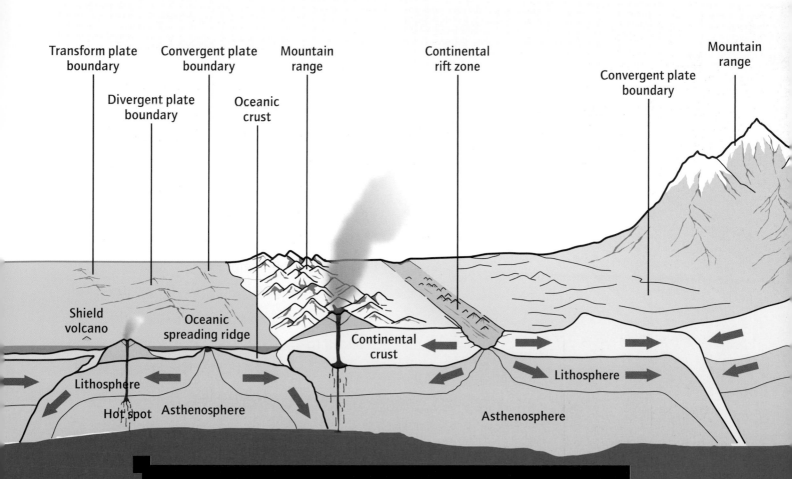

There are several types of plate boundaries. Convergent boundaries either slam into each other to build mountains or slide one over another, creating volcanoes. Divergent boundaries are where two plates push apart, such as the Mid-Atlantic Ridge and the Great Rift Valley in Africa. Transform boundaries grate against each other, like the San Andreas Fault.

EVOLUTION
IN ACTION

The information obtained from a fossil tells us much more than its identity or when it died. Fossil plants and plant parts, such as pollen grains, give us an indication of the vegetation growing at any one time. This, in turn, gives us an indication of the prevailing climate.

Towering forests

In the early **Carboniferous** period there was a marked cooling of Earth, but scientists could not explain what caused this cooling. Dr. Howard Falcon-Lang from the University of Bristol, UK, has an answer. He discovered fossil trees up to 45 m (150 ft.) tall in Newfoundland rocks. They were the ancestors of present day

The pattern of highly sculptured outer walls of morning glory pollen can be recognized by botanists. Identifying pollen can help show when and where a species was present even though the plants themselves may have disappeared.

300 MYA
First **montane forests** appear in the Carboniferous.

150 MYA
Juravenator, a **theropod** dinosaur and bird ancestor without feathers.

147 MYA
Archaeopteryx, a primitive bird with feathers.

C.125 MYA
Microraptor, a dinosaur with feathers.

conifers, but more importantly, they grew on a mountainside. Few fossils exist of upland forests because the conditions on mountains are not good for **fossilization**.

Why was this discovery significant? About 300 million years ago, Europe and North America were joined and straddled the **Equator**. A huge mountain chain ran across the landmass. The Newfoundland rocks were formed in a tiny basin in the heart of these mountains. The surrounding lowlands were covered by tropical forest, but this discovery shows that the mountains were cloaked in forest too – the world's first upland forests. As these new forests grew, the trees converted more carbon dioxide into oxygen, which reduced the greenhouse effect and lowered global temperatures. So, according to Dr. Falcon-Lang, the first mountain forests contributed to climate change.

Evidence from animals

Fossil animals are also indicators of climate change. Professor Jim Marshall from the University of Liverpool, UK, studies the fossils of tiny midges found in **sediments** from a lake in Cumbria, UK. Different species appear when the climate warms or cools. Using the midges as evidence, Marshall has identified two periods of sudden cold in the UK at around 9,000 and 8,000 years ago. This research

The presence of woolly rhinoceros remains in Staffordshire, northwest England, indicates that conditions there were once similar to those on the Arctic tundra today.

shows that the climate of the British Isles is not as stable as it seems.

The most likely explanation for the cold spells is that the **Gulf Stream**, which helps keep Britain mild and wet, slowed down. A period of global warming melted the polar ice caps and changed ocean currents, including the **Atlantic Conveyor**. This works when cold water from the poles sinks to the bottom of the North Atlantic and moves towards the tropics. At the tropics it is replaced by warm surface waters, such as the Gulf Stream, which move from the tropics to the poles. Global warming stopped both the Atlantic Conveyor and the Gulf Stream. Unlike the rest of the world, which warmed up, Britain cooled down. The current episode of global warming could have the same effect.

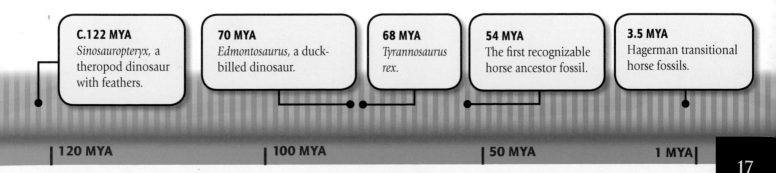

C.122 MYA
Sinosauropteryx, a theropod dinosaur with feathers.

70 MYA
Edmontosaurus, a duck-billed dinosaur.

68 MYA
Tyrannosaurus rex.

54 MYA
The first recognizable horse ancestor fossil.

3.5 MYA
Hagerman transitional horse fossils.

| 120 MYA

| 100 MYA

| 50 MYA

1 MYA|

17

Complexity: the origin of feathers

Fossils can show how complex features might have arisen in **organisms**. They can show how a feature that was designed for one purpose became useful for another purpose. Feathers, for example, have several purposes in modern birds: they help the animal regulate its body temperature, enable it to fly, and are also used to impress a partner. Scientists, however, cannot agree which function evolved first. One argument is that feathers were first utilized as insulation for warm-blooded dinosaurs, but they also became useful for flying. Another view is that birds evolved feathers specifically to fly. Evidence from the **fossil record** is mixed.

Feathered and non-feathered dinosaurs

Fossils from Germany show a 150 million-year-old *Juravenator* dinosaur without feathers. A recent fossil discovery from China, however, shows details of **filamentous** thread-like structures on the skin of *Sinosauropteryx*, a dinosaur that lived around 122 million years ago. They resemble the **contour feathers** of a bird, and could well have helped to regulate the dinosaur's body temperature.

Long feathers

Fossils of *Caudipteryx* and *Microraptor* from China show that these two dinosaurs possessed large feathers. Feathers from both species had a stiff central vein much like those found on modern flight feathers, but these were set in a fan upon the tail or along the forearm. Both dinosaurs were long-legged, short-armed animals that are unlikely to have flown.

Caudipteryx was a peacock-sized dinosaur. It is thought to have had elaborate feathers like those of modern birds. Their function was probably for either insulation or display, and not for flying.

This fossil of a *Microraptor* is about 130 million years old. It lived in a forest that is now Liaoning Province, China.

PILTDOWN CHICKEN

In 1953 a human skull and jawbone that had been unearthed in 1912 at Piltdown, England, was found to be a hoax – the infamous **"Piltdown man"**. In the 1990s history repeated itself, this time with dinosaur fossils. A fossil of a feathered dinosaur was reported to have been found in China. In a scientific paper it was named *Archaeoraptor*, but shortly after publication the scientist involved realized that he had been tricked. This fossil had been "acquired" rather than excavated by the scientist himself. **CT scans** (see page 22) from a special **X-ray** machine revealed the fossil to be a fake, a composite of at least two species. It had the tail and hind limbs of a *Microraptor* dinosaur and the wing parts of a prehistoric fish-eating bird. Since this hoax, scientists have been very careful when checking the source of fossil finds.

One explanation for the fanned feathers is that they helped to streamline the dinosaur's profile when held against the body, thus enabling it to run faster. When held out to one side, these feathers could have acted as air breaks that helped the dinosaur make tighter turns at speed, much like modern ostriches do today. Alternatively, they could have been part of a courtship or threat display that drew attention to the claws on the forelimbs. It is not hard to imagine an arm flapped for a tight turn or a display evolving gradually into a flapping wing with which the dinosaur could take off.

Missing links

When Charles Darwin put forward his theory of evolution in 1858, he predicted that fossil hunters would find creatures that represented transitional stages from one major animal group to the next, such as the animal or animals that led from fish to amphibians or from dinosaurs to birds. At the time he had no examples of these "missing links". Shortly after he published his theory, however, there was a discovery that caused quite a stir.

Early bird

In 1861 the first complete specimen of *Archaeopteryx* was discovered in Germany. It is a dinosaur-like bird that lived about 150 million years ago in the **Jurassic** period.

Archaeopteryx has the wings of a bird and clearly looks like a bird, but its teeth, three fingers with claws, and long bony tail are features shared with small **theropod** dinosaurs. This was the first transitional fossil to be found, and as such provided tangible evidence that supported the theory of evolution.

Ten other specimens of *Archaeopteryx* have since been excavated, and they are so well fossilized that the impressions of flight feathers can be seen in the rock. The feathers are advanced in structure, which means that feathers must have been evolving for some time. Now the search is on for a transition fossil that shows the origins of those feathers.

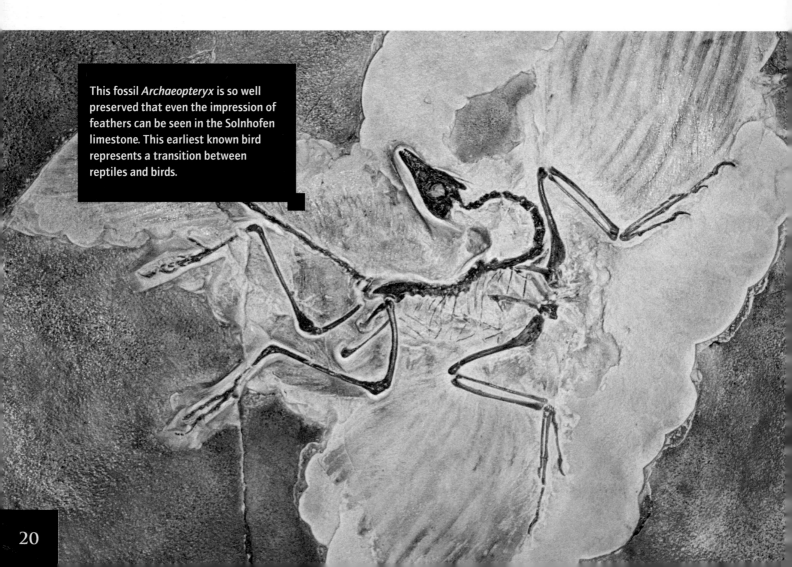

This fossil *Archaeopteryx* is so well preserved that even the impression of feathers can be seen in the Solnhofen limestone. This earliest known bird represents a transition between reptiles and birds.

WITHER THE HORSE

The discovery of a series of transitional fossils enabled palaeontologists to reconstruct the evolution of the horse, currently the most complete and complex evolutionary tree worked out for any animal. There were many species, some of which did not lead to modern horses, but there were several important species in horse evolution. One of the earliest fossils was *Hyracotherium*, a fox-sized creature. It lived 54 million years ago in North America. Its limbs were long, the first sign of an adaptation that allowed the animal to run long distances. Later, about 50 million years ago, came *Orohippus*. It was slimmer and had longer limbs, making it a good jumper.

About 40 million years ago, North America's climate changed from wet to dry, and grasses evolved. The early horses adapted to a drier climate and the arrival of grasses. *Mesohippus* lived on those grass plains. It had long legs to run from predators. *Miohippus*, the species that followed, was larger than its predecessor, but smaller than modern horses. Then *Parahippus*, the size of a small pony, was followed about 20 million years ago by *Merychippus*, which grazed on tough steppe grasses.

By 15 million years ago, many true horses evolved and one line led to *Plesippus*, an ancestral form that is believed to have been the forerunner of modern horses, *Equus*. In 1928 some horse fossils were found in Idaho, USA, dated to about 3.5 million years old. They came from a horse roughly the same size as a modern Arab horse. Called the Hagerman horse, this represents a transitional form leading towards the modern horse.

▼ *A diversity of fossils representing many stages in the development of the horse means that it is the best-documented evolutionary story. Even so, scientists disagree about the way horse evolution occurred.*

1. *Hyracotherium* (oldest fossils)
2. *Mesohippus*
3. *Miohippus*
4. *Merychippus*
5. *Plesippus*
6. *Equus*

A scientist places a fossil skull of an ancestor of modern otters in a CT scanner. It has been adapted to scan objects in microscopic detail using X-rays.

Fossils and disease

Fossil bones and teeth can tell scientists all manner of things about an ancient creature: its size and shape; whether it ran, swam or flew; and what it ate. With modern techniques they can even work out what diseases it had. Much of the information comes from bone, which is a living tissue that is continually formed and re-formed throughout an animal's life. Duck-billed dinosaurs, for example, show an unusually high prevalence of broken and healed bones, including fractured jaws and pelvises. These animals must have skulked away to the forest to hide, or stayed at the centre of the herd while their injuries repaired, surviving on their fat reserves. Many duck-billed dinosaurs have damaged tails, with the subsequent fusing of the bones having put in a distinct kink.

Dinosaur tumours

Bruce Rothschild from Northeastern Ohio University, USA, scanned 10,000 dinosaur fossils using a portable **X-ray** machine and found that some duck-billed dinosaurs had tumours. He found evidence of 29 tumours, most of them benign (harmless), but in a specimen of *Edmontosaurus* he found evidence of a malignant (harmful) tumour. Similarly, in 2003, Peter Larson of Black Hills Institute, South Dakota, USA, found a large brain tumour in the fossil skeleton of a female *Gorgosaurus*, a close relative of *Tyrannosaurus rex*. The tumour is so big that it probably put pressure on the animal's brain. There is also evidence that she fell over often because her sense of balance was impaired. She had a smashed shoulder blade, an infection in the lower jaw, a torn tendon in the leg, and a badly fractured right **fibula** that had not healed. It was the broken leg that probably led to her eventual death.

A cornucopia of dinosaur diseases

Only about one percent of known diseases are evident by looking at bones, so we have no knowledge of 99 percent of dinosaur diseases. Mosquitoes, midges, and ticks undoubtedly attacked them, because several insect species with fully developed piercing mouthparts have been found in amber. Some insects would have carried diseases, just as they do today. Footprints in fossilized trackways have shown dinosaurs with toes missing, and even a dinosaur with a limp. Teeth are often fractured or broken off. Dinosaurs, however, rarely suffered tooth decay. Their teeth were replaced continually so they were unlikely to have developed dental caries (infection and cavities) before they dropped out.

ANCIENT TUBERCULOSIS

John Kappelman from the University of Texas in Austin, USA, examined fossil skull fragments from an ancestor of modern humans who lived 500,000 years ago in what is now Turkey. He revealed that damage to the bone is consistent with scars caused by a **tuberculosis** bacterium responsible for attacking brain membranes. It is the earliest known sign of tuberculosis in humans.

A researcher uses a scanning electron microscope to study infinitesimally small fossil bacteria in ancient rocks, billions of years old. The rocks and fossils were laid down close to the time life evolved on Earth.

NATURE'S GREAT MOMENTS

Nature's greatest moment, perhaps, was the start of life itself. In one of his letters, Charles Darwin suggested that life originated in a "warm little pond" and he might well have been right. Until recently, deep sea hydrothermal vents in the oceans or saltwater rock pools on the shore have been suggested as the cradles for life. These two places would both have had the chemicals and the energy required, from the sun and the centre of Earth respectively, to create life.

However, in 2002 Charles Apel from the University of California in Santa Cruz, USA, revealed that primitive cell membranes assemble more easily in freshwater than saltwater. Life on Earth might have started in a pond after all. However, as yet there are no fossils to support such a hypothesis, but scientists are now finding specimens that are close to the origins of life. The work in this field is often controversial because the specimens are very old, very small and can be open to several different interpretations.

Earliest fossils

Some of the world's oldest rocks are found in Western Australia and they contain the world's oldest fossils. They are 3.2 billion years old, which means that they were laid down not long after Earth had settled down from its initial formation. These fossils contain signs

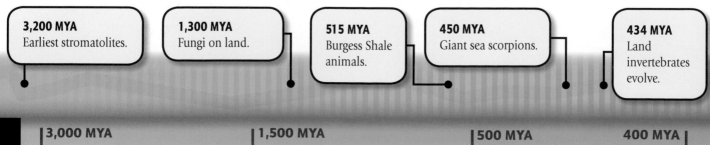

3,200 MYA
Earliest stromatolites.

1,300 MYA
Fungi on land.

515 MYA
Burgess Shale animals.

450 MYA
Giant sea scorpions.

434 MYA
Land invertebrates evolve.

| 3,000 MYA | 1,500 MYA | 500 MYA | 400 MYA |

MYA = million years ago

of life known as **stromatolites**. Stromatolites are formed when **cyanobacteria**, known also as blue-green algae, produce sticky mats to which sand grains stick. The microbes then grow through the grains to form another sticky layer, and so on until a circular, multi-layered reef structure is formed. Modern stromatolites can be seen growing today in Shark Bay, Western Australia. They live in a very salty lagoon where few grazing animals survive, so the stromatolites grow unhindered.

Earliest animals

The early evolution of animals is very difficult to pin down. The first multicellular animals were blobs, like sponges, but they did not stay that way for long.

In 2004 David Bottjer from the University of Southern California, USA, along with co-workers in China, found what they believe to be the earliest bilateral animals – animals in which one side is almost a mirror image of the other side. They were examining Chinese deposits up to 600 million years old.

At first the team found fossils of sponges, but then they came across just ten tiny, oval-shaped animals with a multi-layered body, mouth, gut, and anus. They named them *Vernanimalcula*, meaning "small spring animal", because they were living directly after an extremely long "winter" when ice was

thought to have covered most of the planet on and off for millions of years, a time called "snowball Earth".

At this time only a few simple forms of life survived in the sea. When alive the tiny pioneers probably resembled small, legless **trilobites** and moved by flexing the body. These animals were just a taste of things to come, being in the vanguard of the biggest and most abrupt periods of diversification of life on Earth – the **Cambrian explosion**.

These mounds are stromatolites. They are mats of cyanobacteria and particles of sand that, during the past 4,000 years, have been turned into stone (mineralized). They are to be found at Shark Bay, in Western Australia.

225 MYA
Pterosaurs flying.

125 MYA
Flowering plants.

55 MYA
Bats.

THE CAMBRIAN EXPLOSION

Every **phylum** of animals living today first appeared at the beginning of the Cambrian period about 530 million years ago, a phenomenon known as the "Cambrian explosion". It was one of the biggest periods of diversification in Earth's history. Before the explosion, animal life consisted mainly of single cells or blobs of cells, but afterwards animals appeared that resembled those we see today. At first, however, there were some very strange creatures indeed.

Ediacaran fossils

In the Ediacaran Hills in Australia, the Charnwood Forest in the UK, and the Avalon Peninsula in Canada, fossils have been discovered of creatures that lived 575 million years ago. The oldest is *Charnia*, a 2 m- (6.5 ft.-) long, frond-like animal with branches alternating on either side of a central stem. It lived at the bottom of the sea. Another creature, *Spriggina* is probably an early, long, thin trilobite. *Cloudina* resembles a stack of cones, one inside the other, and *Dickinsonia* is a flat, ovoid creature with tubes emanating from a long central ridge. These animals are so strange that it is difficult to assign them to any major animal group.

Chengjiang fossils

In China's Yunnan Province there is a collection of unusual Cambrian creatures

◀ *This fossil was found in Burgess shale in the Yoho National Park, British Columbia, Canada.*

that lived in the oceans 525 million years ago. The most dramatic is *Anomalocaris*, the smallest known species of which is a 60 cm- (2 ft.-) long, swimming, soft-bodied predator. It has fin-like flaps on each side, two stalked eyes, a disc-like mouth and a pair of grasping arm-like appendages lined with spines in front of the mouth. Larger species were up to 2 m- (6.6 ft.-) long.

The Burgess Shale

The Burgess Shale from British Columbia, Canada, is roughly 515 million years old, and contains **organisms** that would not look out of place in a science fiction movie. *Hallucigenia* is a worm-like creature sprouting many appendages; *Wiwaxia* is an armoured slug-like animal with two rows of upright scales; and *Opabinia* is a segmented animal with a head sporting five stalked eyes and a flexible, hose-like proboscis ending in grasping spines.

▲ *This is a fossil of* Dickinsonia *from late-Ediacaran rocks. It is 13 cm- (5 in.-) long, but nobody is sure what type of animal it may be. Some say it is a flat-bodied worm, others suggest a soft-bodied coral.*

WHY AN EXPLOSION?

The reason for the Cambrian explosion is unclear. One suggestion is that there was more oxygen in the atmosphere at the start of the Cambrian, and therefore more oxygen dissolved in the oceans. Another is that the necessary "genetic tool kit" had evolved to give rise to millions of years of experimentation. The problem is the sparseness of fossils. The organisms may not have had parts that fossilized, and their environment may not have been conducive to **fossilization**. When animals died they simply decayed. It is a fundamental problem of palaeontology, compounded in this case by the great age of the rocks.

Leaving the water

In order to obtain information about the first organisms on land, scientists turned not to fossils but to genetics. Blair Hedges and his team at Penn State University, USA, examined thousands of genes from hundreds of different species of living organisms and found 119 that were common to fungi, plants, and animals. As soon as each new species evolves, mutations in these genes occur at regular intervals. It is like the ticking of a clock – a "molecular clock". With this knowledge, the researchers could backtrack to the origin of each species. They then checked their "clock" against recognized events at known dates in the **fossil record**. From all this information they could work out when each species originated. Their research indicated that land organisms appeared much earlier than anybody had expected.

First land fungi

Hedges and his team discovered that fungi spread across the land roughly 1,300 million years ago, and that plants came along around 700 million years ago. Prior to this there would have been a few bacteria and algae on the barren landscape and little else. The fungi teamed up with algae to form the first **lichens**, and along with the early mosses the Earth's landscape began to be covered with green plants. An unexpected outcome of this research is a plausible explanation for the theory of "snowball Earth" and the Cambrian explosion. The early plants took in carbon dioxide and emitted oxygen. This reduced the greenhouse effect, so the planet cooled, creating "snowball Earth" – when much of Earth, it has been suggested, was covered by ice. The increase in oxygen, however, gave a boost to the evolution of animals under the sea and produced the Cambrian explosion that followed.

Lichens, similar to these lichens from South Africa, were among the first organisms to invade the land. They are the result of a symbiotic relationship between a fungus and an alga.

Organisms resembling liverworts were among the first green plants to invade the land. Fossils similar to liverworts have been found in rocks 475 million years old.

First land animals

Fossilized trackways in 500 million-year-old rocks of ancient sand dunes from a quarry at Kingston, Canada, indicate that one of the earliest animals to leave the water and head for land is a woodlouse-like creature known as an euthycarcinoid. It was an ancestor of insects. The earliest known index fossil of a land animal is of a 1 cm- (0.4 in.-) long millipede-like creature that was discovered in about 428-million-year-old rocks in Scotland. Modern millipedes eat decaying plant material so it is probably no coincidence that among the earliest animals was one that could take advantage of the burgeoning plant life.

EARLY LAND FOSSILS

Some of the earliest fossil evidence of life on land is found in rocks dated 475 million years old from Oman, in the Middle East. Charles Wellman, from the University of Sheffield, UK, examined fossil spores from these rocks and revealed that they were from plants on land rather than algae from the sea. The plants that produced them would have been similar to today's liverworts, and were among the first plants to colonize the land.

This fossilized sea scorpion lived about 400 million years ago. Like modern spiders, to which it is distantly related, it has two main body parts, as well as large paddle-like swimming legs, a pair of pincers, and a spine at the end of the abdomen.

First giants

Throughout the history of life on Earth there has been a tendency for certain **lineages** of evolving organisms to increase in size over time. The benefits of a large body size include more "muscle" to gain territory, food, and a mate; an enhanced ability to fight off predators and rivals; and more fat reserves to survive through lean times. However, the costs of a large body size include the need for more food, a lengthy "growing up" period, fewer offspring, a reduced ability to cope with environmental change, and therefore an increased susceptibility to extinction.

LAND GIANTS

About 300 million years ago giant dragonflies ruled the skies. The largest had a wingspan of over 76 cm (30 in.). It grew this big because of two factors:

1. The composition of the atmosphere;
2. The structure of the insect's **tracheal breathing system.**

We have **blood vessels** and **blood corpuscles** that take oxygen from the lungs to the tissues, but insects have a system of tubes that takes the oxygen directly from the outside to all parts of their body. If the tubing system is too large it will not work, but its maximum size is determined by the amount of oxygen in the atmosphere. The higher the oxygen level, the bigger an insect can grow. At the time when those ancient giant insects lived, lowlands were covered with tropical forests of giant ferns and trees. Photosynthesis was taking place on a large scale, which led to high oxygen levels in the atmosphere. During the **Carboniferous** this enabled giant dragonflies to be the top aerial predators.

Leedsichthys was a monster. It was probably the largest fish that ever lived, yet it was a harmless filter-feeder – like the modern basking shark.

Sea monsters

Eurypterids, or sea scorpions, were early giants. They were aquatic relatives of today's scorpions and spiders, except for the fact that eurypterids were up to 2.5 m (8 ft.) long. In 2007, the claw of such a monster was found in rocks estimated to be 450 million years old from a quarry in Germany. Trilobites grew to large sizes, too. A record-breaking 70 cm (27.5 in.) trilobite was taken from rocks 445 million years old at Hudson Bay, Canada. It probably fed on shrimps and worms on the sea floor. Later in the fossil record, one of the largest known **ammonites**, 2.55 m (8.4 ft.) across, was found in rocks of the late-**Cretaceous** period in Germany.

The largest fish that ever swam in the ocean appeared about 150 million years ago, at the end of the **Jurassic**. It was a monster called *Leedsichthys*. An entire backbone has yet to be found, but it is estimated to be up to 22 m- (72 ft.-) long, which is double the size of the world's largest living fish, the whale shark.

Leedsichthys fed in a similar way to the whale shark, having 40,000 tiny teeth with which it sieved plankton, small fish, shrimps, and jellyfish from the water.

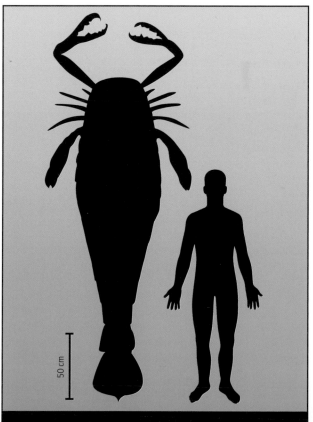

The largest ancient sea scorpions were much longer than a person is tall. They lived in the sea about 450 million years ago.

First flight

Powered flight evolved on at least four separate occasions during Earth's history. Insects were the first creatures to fly, followed by pterosaurs, then birds and bats. This is a case of **convergent evolution**, in which unrelated species have developed the same solution to an environmental challenge.

Insect flight

Insects evolved during the **Silurian** period, about 434 million years ago. The first sign of insect wings was found on a creature called *Rhyniognatha* in rocks from Scotland dated 407–396 million years old. This means that insects took to the air not long after the insect group evolved. Many of the earliest flyers probably fluttered about close to the ground, but the earliest skilled flyers were the predecessors of the dragonflies.

Flying reptiles

The pterosaurs, close relatives of the dinosaurs, were airborne about 225 million years ago. While other reptiles, including some species of dinosaurs, might well have glided, the pterosaur was the first vertebrate to power itself into the air by flapping its

Large pterosaurs most likely soared like modern vultures. They dominated the skies while the dinosaurs ruled Earth. They were the first vertebrates to take to the air.

leathery wings. Some grew to a tremendous size. The largest known pterosaur was *Quetzalcoatlus*, which had a wingspan of 15 m (50 ft.). The pterosaurs disappeared long before their terrestrial cousins during the Cretaceous-**Tertiary** mass extinction about 65.95 million years ago, leaving the skies clear for the evolution of birds.

"Trees down" and "ground up"

It is generally thought that birds evolved directly from dinosaurs, and that they are living dinosaurs. The first bird-like fossil to be found was *Archaeopteryx*, which appeared in the fossil record about 150 million years ago.

How it became airborne is debated hotly. Some scientists believe in the "trees-down" hypothesis in which the bird climbed a tree then glided and flapped back down to earth. Others suggest a "ground-up" hypothesis in which the early bird ran fast, made long hops, and finally flapped its wings to take it into the air. Why it flew at first is unknown – probably to escape enemies or to help it find food, just as birds do today. The largest bird that ever lived was *Argentavis*, a bird of prey with a 7 m- (23 ft.-) wingspan that lived about 6 million years ago.

FLYING MAMMALS

The first and only mammals to achieve powered flight naturally are the bats. They were aloft about 55 million years ago during the **Eocene**. The earliest known bats included *Icaronycteris* with a wingspan of about 30 cm (12 in.). By this time they were already active flyers and anatomically very similar to modern bats. Earlier fossils, when bats were evolving from non-flying mammals, have yet to be found.

An abominable mystery: flowering plants

Charles Darwin referred to the origin of flowering plants as "an abominable mystery". The problem is that flowering plants, or **angiosperms**, appear very suddenly in the fossil record. There is rapid diversification from the mid-Cretaceous (a plant equivalent of the Cambrian explosion), but there are no obvious ancestors in the preceding 100 million years. The earliest known fossil flowers include *Archaefructus* from China, dated at 142–125 million years ago, and a 25 cm- (10 in.-) high wetland herb with fern-like leaves and tiny flowers named *Bevhalstia pebja*, which is found in 130 million-year-old mudstones from a quarry to the south of London, UK. There are two theories suggesting how these flowering plants might have arisen.

Woody theory

James A. Doyle from the University of California and Michael J. Donoghue from the University of Arizona, USA, believe that the first angiosperms were small- to medium-sized woody trees with broad leaves and large flowers. They had a slow life cycle. These plants would have been similar to today's magnolia, bay laurel, avocado, and cinnamon.

Herb theory

David W. Taylor from Indiana University Southeast and Leo J. Hickey from Yale University, USA, believe that the earliest flowering plants were tropical herbaceous plants called paleoherbs, with uncomplicated flowers and a rapid life cycle. They would have been primitive versions of today's birthworts, lotus, and waterlilies.

One theory suggests that plants related to the group that includes modern water lilies were the first flowering plants. Fossil seeds of water lilies have been found in late-Cretaceous rocks, and one of the earliest known flowers found in lower-Cretaceous rocks is thought to have been aquatic.

Fossils of the earliest flowering plants are notoriously difficult to find. Later fossils, like this fossil sycamore leaf, are found more frequently.

Climate change

It is likely that climate change triggered the rise of the flowering plants. They differed from earlier plants in having a seed contained in a fruit rather than a naked seed like ferns and conifers. The first angiosperms were probably pollinated by insects, such as beetles that chomped their way into the pollen-bearing parts of the flower. Over millions of years the relationship became much closer, with some species of flowering plants depending on a particular species of insect for pollination.

The arrival of flowering plants is also linked to the rise of the large, low browsing and grazing dinosaurs, such as the duck-billed dinosaurs. They found the newly evolving plants nutritious and plentiful. In fact, the flowering plants and plant-eating dinosaurs could also have **co-evolved**. Scientists have suggested a link between the newly evolving plant species and the evolution of complex jaw mechanisms in certain dinosaurs.

PRIMITIVE FLOWER

Molecular biologist Elizabeth Zimmer from the Smithsonian Institution, USA, has been studying the **DNA** from a range of living flowering plants. She has identified the oldest surviving lineage as being the Amborellaceae, which goes back 130 million years. It is a family with only one living representative, *Amborella trichopoda*, which grows on New Caledonia in the Pacific Ocean. The small woody plant is described as a "living fossil".

WE CAME FROM THE SEA

Evidence shows that the vertebrates (animals with backbones) made their first forays onto land during the **Devonian** period (416–359.2 million years ago). John Long, from Museum Victoria in Melbourne, and colleagues at Monash University, made a significant find in Western Australia that helps to tell the story. They found fossils of a **lobe-finned** fish called *Gogonasus* in rocks 380 million years old. It had large holes in its skull for breathing through the top of its head and muscular front fins with bones that would eventually become the radius and ulna, the same bones as those in the human forearm. Its fins would have supported its body out of water. *Gogonasus* is one of several types of lobe-finned fish that were emerging from warm, oxygen-poor, shallow-water habitats.

"Fishapod" – half fish, half amphibian

Tiktaalik is another transitional fossil. It was found in 2006 on Ellesmere Island, in the Canadian Arctic, by a team led by Neil Shubin from the University of Chicago, USA, and Ted Daeschler from the Academy of Natural Sciences, Philadelphia, USA. The rocks it was found in are 375 million years old. The creature itself is a fish with gills and fish scales, but its head resembles that of a crocodile. Its fins have fin rays, like the fins of fish, but also sturdy wrist

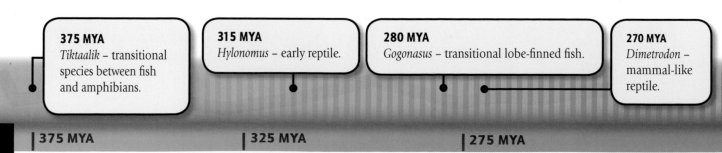

375 MYA
Tiktaalik – transitional species between fish and amphibians.

315 MYA
Hylonomus – early reptile.

280 MYA
Gogonasus – transitional lobe-finned fish.

270 MYA
Dimetrodon – mammal-like reptile.

| 375 MYA | 325 MYA | 275 MYA |

MYA = million years ago

COLOURFUL WORLD

When fish made the transition from water to land, scientists believe that they viewed their new environment in colour. Fish can see some colours, mostly reds and blues, but Helena Bailes from the University of Queensland, Australia, examined the eyes of lungfishes and found that they were similar to the eyes of land vertebrates. Lungfish have sensory cells in the retina of the eye that are capable of detecting more colours than other species of fish. The significance of this is that lungfish are close relatives of the ancient lobe-finned fishes that left the water millions of years ago. One lungfish in particular was interesting. It is called *Neoceratodus* and has changed very little for about 135 million years. By studying its eyes and comparing them to the eyes of living land vertebrates, scientists can tell what the amphibian-like creatures saw when they first stepped out onto the land. So way back in the Devonian, the images they saw were probably in colour.

bones. *Tiktaalik's* gills have no gill covers so it would have had a moveable neck – modern fish cannot move their necks. Its ribs are thick, like those of a land vertebrate. These features make *Tiktaalik* a clear transitional form between fish and amphibians. It could prop up its body and push its head out of the shallow-water environment in which it lived. Maybe this was to search for new sources of food.

Tiktaalik was not, however, pushing its way deliberately onto land. It had merely acquired the necessary adaptations to do so. Later transitional forms were to gradually make that journey. The discovery of *Tiktaalik* was expected, being an interesting example of how scientists predict what some creatures might have looked like. Scientists knew that a creature like this must have existed in order to explain a gap in the evolution of land animals. So they went in search of it in rocks of the right age, with the find confirming the prediction that the scientists had made.

The Australian lungfish sees in colour. Creatures similar to lungfishes were probably the first vertebrates to clamber onto the land. They must also have seen in colour.

230 MYA
Eoraptor – earliest dinosaur.

210 MYA
Antetonitrus – early **sauropod** (lizard-hipped dinosaur).

200 MYA
Eocursor – early herbivorous **ornithischian** (bird-hipped) dinosaur.

C.80 MYA
Giganotosaurus – giant meat-eating dinosaur.

| 225 MYA | 175 MYA | 125 MYA | 75 MYA |

Age of the dinosaurs

The dinosaurs are often depicted as huge bumbling creatures that died out suddenly, but they were, in fact, one of the most successful groups of animals that ever lived on Earth. Dinosaurs were descended from reptiles and had humble beginnings, but existed for 165 million years and dominated the Earth for 150 million years.

Hylonomus, the earliest known fossil reptile, was just 20–30 cm- (7–12 in.-) long. These early reptiles evolved in the shadow of large amphibians in the **Carboniferous** swamp forests, but a mini-Ice Age changed all that. The amphibians went into decline and the reptiles began to get bigger.

The first large reptiles were not dinosaurs. The 3 m- (10 ft.-) long *Dimetrodon* was a **pelycosaur** that lived in the **Permian**. A huge fin-like sail on its back helped it to warm up or cool down, and it was generally more like a mammal than a reptile. The group of reptiles that gave rise to the dinosaurs were the **thecodonts**, crocodile-like creatures.

First dinosaurs

The earliest dinosaur fossils are from 230 million-year-old rocks in Argentina, Brazil, and Madagascar. *Eoraptor*, a small meat-eating dinosaur, is considered the most primitive, but it has some of the advanced features associated with dinosaurs so there must be even older specimens waiting to be discovered.

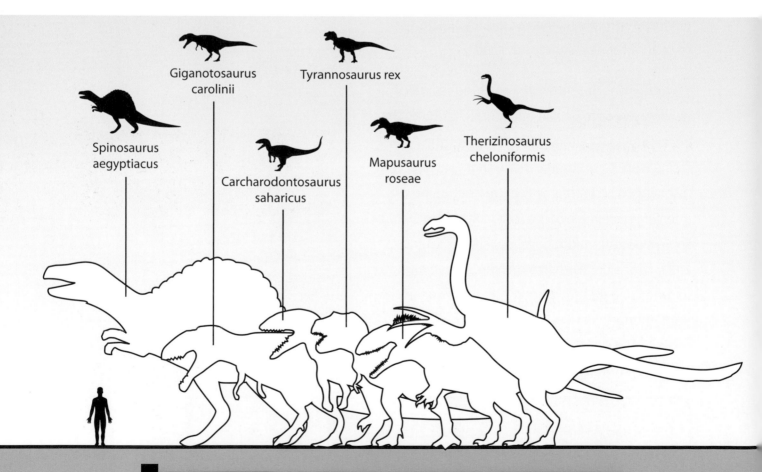

Spinosaurus aegyptiacus

Giganotosaurus carolinii

Tyrannosaurus rex

Carcharodontosaurus saharicus

Mapusaurus roseae

Therizinosaurus cheloniformis

Dinosaurs, like all other groups of animals before and since them, were prone to grow big. *Tyrannosaurus rex* was puny compared to its close meat-eating relatives.

Two dinosaur groups

Palaeontologists divide the dinosaurs into two main groups. The **Ornithischia**, or bird-hipped dinosaurs, were mainly beaked herbivores. The **Saurischia**, or lizard-hipped dinosaurs, includes **theropods**, the carnivorous dinosaurs, and giant **sauropods**.

One of the earliest ornithischians is *Eocursur*, a small plant-eating dinosaur less than 1 m- (3.3 ft.-) long that ran on two legs roughly 200 million years ago. Its descendents included such well-known giants as the upright *Iguanodon* (one of the first dinosaur fossils to be found), the rhino-like *Triceratops*, and the *Stegosaurus* with its two rows of upright, bony plates on its back.

With a brain the size of a walnut, *Stegosaurus* is likely to have been the dumbest dinosaur. It had a nerve centre, like a second brain, in its hips in order to move.

Despite the name, birds evolved from lizard-hipped dinosaurs rather than bird-hipped dinosaurs. They evolved the bird-like pelvis quite separately from the bird-hipped dinosaurs, a case of **convergent evolution**.

Karen Chin surveys her collection of dinosaur droppings, known as coprolites, at the University of California, Santa Barbara, USA.

LIVING ALONGSIDE DINOSAURS

In the same period that the dinosaurs were evolving, during the **Triassic**, the first turtles, herbivorous **aetosaurs**, and crocodile-like **phytosaurs** appeared, too. It was also the dawn of the mammals, although they were no more than tiny insect- and seed-eaters at the time. Of the insects, there is evidence from fossilized dinosaur dung, known as coprolites, that scarab beetles were among the creatures clearing up dinosaur waste. Plants at first included the conifers, cycads, bennettitales, and ferns, and they were followed later by the flowering plants.

Herds and packs

Large numbers of footprints with varying track lengths on dinosaur trackways show that many dinosaurs of different sizes sometimes travelled together. Sauropods travelled in herds, with youngsters walking in the centre of the herd for safety, like baby elephants do today. Large numbers of fossils of a single species found in the same place, known as "bonebeds", are further evidence of herding behaviour. Bonebeds containing the fossils of *Styracosaurus*, a dinosaur with a six-pointed frill behind its head, and *Protoceratops*, with a simple frill, have been found.

In China, a group of baby *Psittacosaurus* was discovered together, perhaps indicating that they had been in a "dinosaur crèche". There is evidence that some dinosaurs had nesting colonies – a behaviour they must have passed on to birds. The duck-billed *Maiasaura* was the first dinosaur to be discovered alongside nests containing eggs and young. The nests were hollows in the ground about 2 m (6.6 ft.) in diameter and 9 m (30 ft.) apart. Each nest contained about 25 round, grapefruit-sized eggs. In Montana there are 40 nests on a site that would have been an island in the **Cretaceous**. It would have been a noisy place, too. The crest on the top of a duck-billed dinosaur's head is thought to have been a resonator that helped amplify the animal's calls. *Parasaurolophus* has a particularly complicated system linking its crest to its nasal passages, much like a trombone.

Some of the small and medium sized meat-eaters probably hunted in packs. An example is *Velociraptor*, which killed its prey with an enormous sickle-like claw on its hind feet. A whole pack of these predators would rip large prey to shreds. A remarkable fossil from the Gobi Desert, discovered in 1971, shows the dramatic moment when a *Velociraptor's* foot claw slices into the neck of a *Protoceratops*. But even a ferocious *Velociraptor* had to be careful how it hunted – the herbivore appears to have bitten and broken the predator's arm.

The duck-billed dinosaurs lived in large herds, looked after their young, and some went on extensive migrations.

These fossils show how a young, predatory *Velociraptor* attacked a plant-eating *Protoceratops*, but died itself. The *Protoceratops* bit into the *Velociraptor's* right hand with its beak-like jaws, while the predator's hind claw was embedded in the *Protoceratops'* belly. Both died and have been fossilized.

Dinosaur trackways give scientists information about what type of dinosaur was on the move, how big it was, how fast it was moving and where it was going. These dinosaur footprints are from the Cretaceous period and can be seen on this uplifted, fossilized lake bed in Sucre, Bolivia.

FAST AND SLOW

Dinosaur footprints in fossilized trackways give an indication of a dinosaur's speed. The animal would probably not have been running fast over the wet sand in which it left its print, but an informed guess at its shape, size and species can be made. It is thought that a medium-sized predator, such as *Velociraptor*, would have had a top speed of about 44 kmh (27 mph), faster than an Olympic sprinter! Comparing dinosaurs with living animals might also give us an indication of the speed in which they travelled. For instance, the giant sauropods, such as *Brachiosaurus*, moved in a similar manner to elephants.

Giant of giants

One of the first sauropods was *Antetonitrus*, which lived about 210 million years ago. Fossils of *Antetonitrus* discovered in South Africa show the first signs of thick limbs like the legs of an elephant. Scientists call this type of movement on four thick legs "graviportal quadrapedalism". It means that the body's weight is supported on four thick legs, like columns. This type of movement enabled the sauropods to develop into the largest animals ever to have walked on Earth. In Argentina, new species of record-sized huge sauropods are being found all the time. One of the biggest is *Argentinosaurus*, which lived during the Cretaceous. It was 45 m (150 ft.) long and weighed an estimated 100 metric tons.

Ultimate land predator

One of the largest known meat-eating dinosaurs is *Giganotosaurus*, which lived about 80 million years ago. It was up to 14.3 m (47 ft.) long and weighed about 7 metric tons, a little bigger than the 14 m- (46 ft.-) long *Tyrannosaurus rex*. There is currently a debate about whether these giants were predators or scavengers. Even larger was *Spinosaurus*, an 18 m- (59 ft.-) long meat-eater. It is recognized by its elongated head and 2 m- (6.6 ft.-) long vertebral spines that either formed a sail, possibly for controlling its body temperature (called thermoregulation), or were covered in muscles to form a large hump.

WARM- OR COLD-BLOODED?

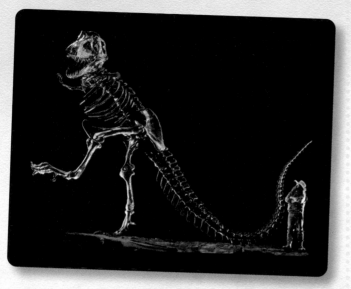

▲ *Museums made their dinosaur skeleton's more active and animated after the revelation that dinosaurs might have been warm-blooded.*

Did you know humans are warm-blooded, or **endothermic**? We can control our body temperature within certain limits. Most reptiles are cold-blooded, or **ectothermic**. Their body temperature is directly dependent on the environment. U.S. palaeontologists Robert Bakker and John Ostrom, however, argued that dinosaurs were successful because they were endothermic, being able to generate internal heat to regulate their body temperature like birds and mammals.

Some dinosaurs were fast runners so must have had high **metabolic rates**, which means that their body burned calories rapidly. The fastest known dinosaur was *Ornithomimus*, which is thought to have reached 64 kmh (40 mph). Theropod and **ornithopod** dinosaurs also have a larger than average brain size, a feature of endothermic animals. The largest brain to body size is found in 3.5 m- (12 ft.-) long troodontids, believed to have been the smartest dinosaurs – roughly as intelligent as modern birds, which are much smaller with smaller brains.

Some dinosaurs lived near Earth's poles. Fossils have been found on the north slope of Alaska, where it was seasonally cooler during the late-Cretaceous. Many probably migrated south in winter, but evidence suggests some remained there all year. This also indicates that dinosaurs had the ability to maintain a comfortable body temperature, thus enabling them to live through the seasonally cooler periods.

There is evidence against the warm-blooded hypothesis of Bakker and Ostrom. Large dinosaurs could have warmed up and cooled down slowly, thereby maintaining a comfortable average body temperature. The global climate was also warmer at that time, although there were fluctuations from season to season. Dinosaurs were generally scaly, a feature of ectotherms, although many theropods are now thought to have had feathers, an indication of endothermy.

The most compelling evidence against the warm-blooded theory, however, is that endotherms tend to have folded bones in the **sinuses**, an adaptation that restricts water loss from the body. Dinosaurs do not have this. This continuing debate is unlikely to end.

Ichthyosaurs ruled the seas from 230 million years ago, long before the dinosaurs evolved on land. They died out 25 million years before dinosaurs went extinct, ousted by the plesiosaurs.

Return to the sea

On several occasions in Earth's history, land vertebrates have returned to a life in the ocean.

Long- and short-necks

Long-necked plesiosaurs and short-necked pliosaurs looked like dinosaurs with paddles. Many were slow-swimming fish and squid eaters, but some grew into sea monsters. One species, *Liopleurodon*, was about 24 m (80 ft.) long and, during the **Jurassic**, it had the largest jaws in the sea. It ate just about anything and was at the top of the food chain. Some people have suggested that a plesiosaur could have survived in Loch Ness and is the monster reported to have been seen there. But this is unlikely because there is not enough food in Loch Ness to satisfy a hungry plesiosaur, let alone an entire family.

Dolphin-like reptiles

During the Triassic, not long after the reptiles were evolving into giants on land, one branch of reptiles gave rise to the ichthyosaurs. They were air breathing, dolphin-like reptiles that dominated the seas. The smallest ichthyosaur was the length of your arm and the largest was 15 m (50 ft.) long. The earliest known ichthyosaur is *Utatsusaurus*, a lizard with flippers. To swim it undulated its body like an eel. Later ichthyosaurs swam more like modern sharks, which use their tail fin. They also had a similar body plan to sharks and dolphins, an example of convergent evolution.

First whales

For many years the evolution of cetaceans (whales and dolphins) was uncertain because so few fossils had been found. However, a series of transitional fossils excavated in Pakistan and Egypt is beginning to make

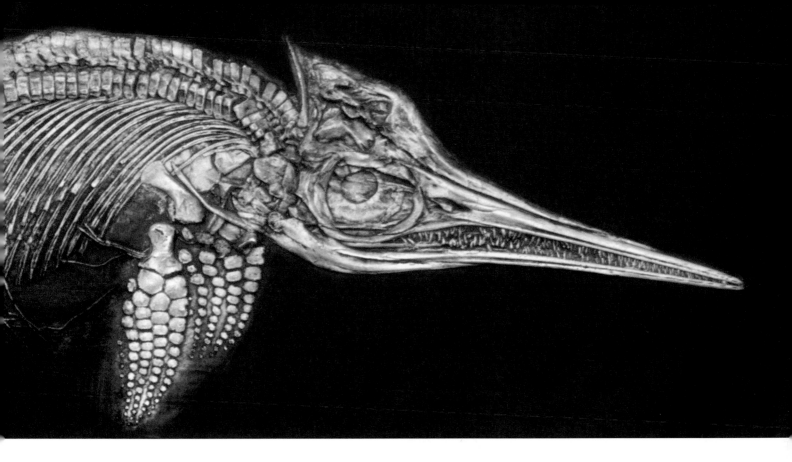

the picture clearer. Cetaceans descended from land mammals related to ancient hippopotamuses. *Pakicetus*, a dog-like meat-eater with hoofed feet, is one of the earliest. It lived about 52 million years ago and looked nothing like a whale (or a hippo for that matter), but it shared the same unusual ear structure with whales and dolphins. The next important find was *Ambulocetus*, which lived 50 million years ago. It is 3 m (10 ft.) long, and its shape suggests an amphibious creature. *Ambulocetus* had a long snout, like a crocodile, and probably hunted like one, intercepting prey in the shallows. The next stage is represented by *Rodhocetus*, which lived about 47 million years ago. It looked more like a modern whale, and had large paddle-like forelimbs with which it could have supported itself on land, in much the same way that sea lions do today. *Rodhocetus* was probably amphibious.

The most whale-like of these early cetaceans is the next stage – the 18 m- (60 ft.-) long *Basilosaurus*. It lived entirely in the sea about 30–45 million years ago. The link to the land is still evident, however, for *Basilosaurus* had tiny hind limbs still visible outside the body. These early whales gave rise to many whale and dolphin species during the **Miocene**, some of which resemble the cetaceans we see in the oceans today.

A *Megalodon* shark tooth (left) is compared to a modern great white shark tooth (right). *Megalodon* was monstrous, about 17 m (56 ft.) long. It fed on the many species of whales that were evolving between 15 and 1.5 million years ago.

A skull of *Homo erectus*, an example of the first early humans to have left Africa. It was found to the southwest of Tbilisi, the capital of Georgia.

First humans

About 5 million years ago, fossil evidence shows that the human **lineage** split from the chimpanzees and our earliest ancestors evolved into the Australopithecines, which means "southern ape-men". Possibly the most famous Australopithecine fossil was "Lucy", excavated in Ethiopia in 1974, she lived 3.2 million years ago. Lucy had a small brain and long arms. She was a primate that walked upright – one of the first defining characteristics of humans.

By about 3 million years ago the Australopithecines were mostly gone. Global temperatures had dropped and most of Africa slowly dried out. Instead of eating plentiful forest fruits the "ape-men" had to survive on meagre pickings from the growing areas of open grassland. Several types of ape-men adapted to the changing conditions. *Paranthropus boisei* developed enormous jaws and facial muscles to chew on tough vegetation, such as roots and tubers. Small-toothed *Homo habilis* ate anything it could find, including scavenging at predator kills.

The adaptation to eating meat was important. Meat is energy-rich and requires a shorter gut, so energy was directed elsewhere in the body – to the brain. With more meat in the diet, more energy could be diverted to brain development. About 2 million years ago these human-like ape-men – called hominids – were intelligent enough to use rocks to smash bones for the marrow inside, which contains the fatty acids needed for brain growth. They were eating the right foods for brain development. As the Ice Age took a hold on the planet, easy-going *Paranthropus boisei* died out, but the bigger brained *Homo habilis* survived by using its intelligence to help it adapt to a changing world.

Early humans

Homo ergaster lived around 2 million years ago. As Africa dried out, *Homo ergaster* was one of the first hominids to regulate its body temperature by sweating. It walked long distances to find food, developing a narrower pelvis to increase the efficiency of its walking muscles. This narrow pelvis restricted childbirth, so babies had to be less well developed at birth. As a result of this, offspring had to be looked after for longer, which led to the start of what we now call the nuclear family. Nuclear families consist of parents and offspring.

About 1.9 million years ago, our human ancestors either moved out of Africa and reached other continents or were evolving elsewhere. Fossils of *Homo erectus*, the next hominid stage, have been found in southern Asia and Europe.

Modern humans

Modern humans – *Homo sapiens* – evolved roughly 250,000 years ago, but about 100,000 years ago our ancestors were almost wiped out by a series of environmental catastrophes. An extensive drought, coupled with the aftermath of the Toba volcano eruption on Sumatra 70,000 years ago, almost killed off our species before it had really got going.

Climate studies show that the global temperature dropped, creating an instant Ice Age that lasted for 1,400 years. In Africa, for example, the temperature dropped by 9°C (16°F). **DNA** studies reveal that probably no more than 10,000 modern people survived.

This caused a genetic bottleneck, the reason for the limited genetic diversity seen in humans today. Forced by the severe conditions these early people struck out for pastures new, reaching the Middle East by 90,000 years ago, Australia by 50,000 years ago, Europe by 40,000 years ago, and North America by about 11,000 years ago.

GENETIC BOTTLENECK

Genetic bottleneck occurs when a significant percentage (more than 50 percent) of a population is killed or prevented from reproducing. Survivors breed amongst themselves but the small number of possible mates results in a smaller variety of genes in successive generations. This means that if a disease hits the population, or the climate changes, fewer individuals have genes that might help them adapt and survive. In other words, if one is hit, all are wiped out.

This is the skull of a four-year-old "ape-man". Known as the Taung child, he lived 2 million years ago. The skull was found in South Africa. The child was probably killed by a large bird of prey.

TIMELINE

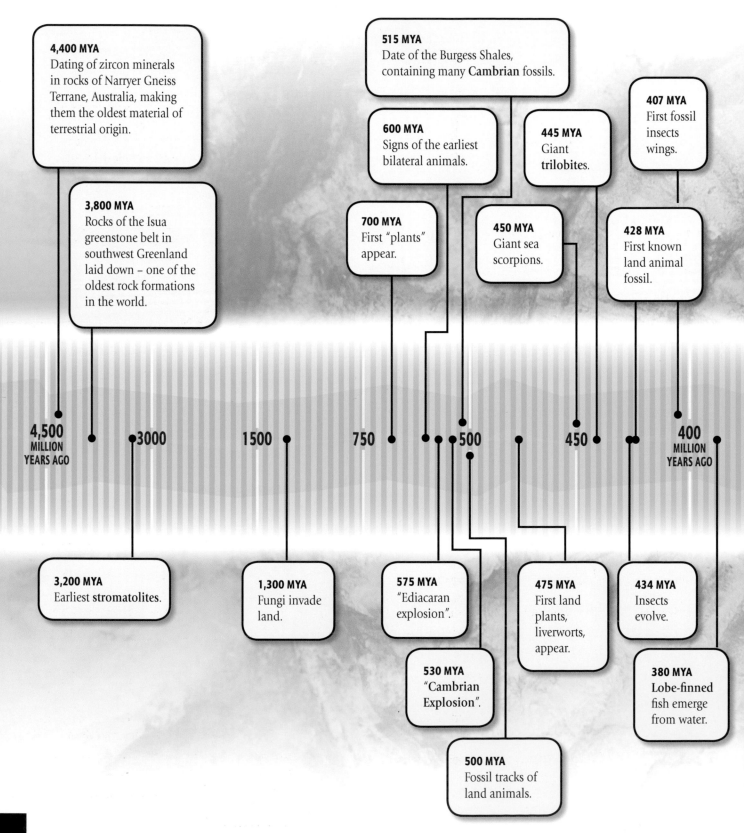

4,400 MYA
Dating of zircon minerals in rocks of Narryer Gneiss Terrane, Australia, making them the oldest material of terrestrial origin.

3,800 MYA
Rocks of the Isua greenstone belt in southwest Greenland laid down – one of the oldest rock formations in the world.

515 MYA
Date of the Burgess Shales, containing many **Cambrian** fossils.

407 MYA
First fossil insects wings.

600 MYA
Signs of the earliest bilateral animals.

445 MYA
Giant **trilobites**.

700 MYA
First "plants" appear.

450 MYA
Giant sea scorpions.

428 MYA
First known land animal fossil.

4,500
MILLION
YEARS AGO

3000

1500

750

500

450

400
MILLION
YEARS AGO

3,200 MYA
Earliest **stromatolites**.

1,300 MYA
Fungi invade land.

575 MYA
"Ediacaran explosion".

475 MYA
First land plants, liverworts, appear.

434 MYA
Insects evolve.

530 MYA
"Cambrian Explosion".

380 MYA
Lobe-finned fish emerge from water.

500 MYA
Fossil tracks of land animals.

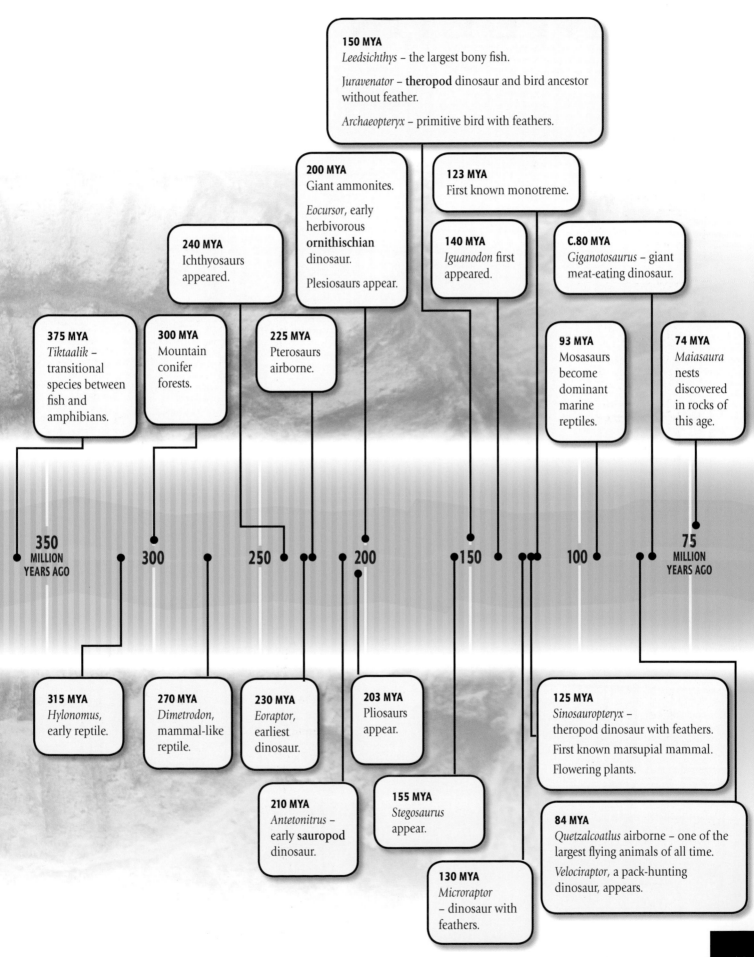

150 MYA
Leedsichthys – the largest bony fish.

Juravenator – **theropod** dinosaur and bird ancestor without feather.

Archaeopteryx – primitive bird with feathers.

200 MYA
Giant ammonites.

Eocursor, early herbivorous **ornithischian** dinosaur.

Plesiosaurs appear.

123 MYA
First known monotreme.

240 MYA
Ichthyosaurs appeared.

140 MYA
Iguanodon first appeared.

C.80 MYA
Giganotosaurus – giant meat-eating dinosaur.

375 MYA
Tiktaalik – transitional species between fish and amphibians.

300 MYA
Mountain conifer forests.

225 MYA
Pterosaurs airborne.

93 MYA
Mosasaurs become dominant marine reptiles.

74 MYA
Maiasaura nests discovered in rocks of this age.

350 MILLION YEARS AGO

300

250

200

150

100

75 MILLION YEARS AGO

315 MYA
Hylonomus, early reptile.

270 MYA
Dimetrodon, mammal-like reptile.

230 MYA
Eoraptor, earliest dinosaur.

203 MYA
Pliosaurs appear.

125 MYA
Sinosauropteryx – theropod dinosaur with feathers.

First known marsupial mammal.

Flowering plants.

210 MYA
Antetonitrus – early **sauropod** dinosaur.

155 MYA
Stegosaurus appear.

84 MYA
Quetzalcoatlus airborne – one of the largest flying animals of all time.

Velociraptor, a pack-hunting dinosaur, appears.

130 MYA
Microraptor – dinosaur with feathers.

49

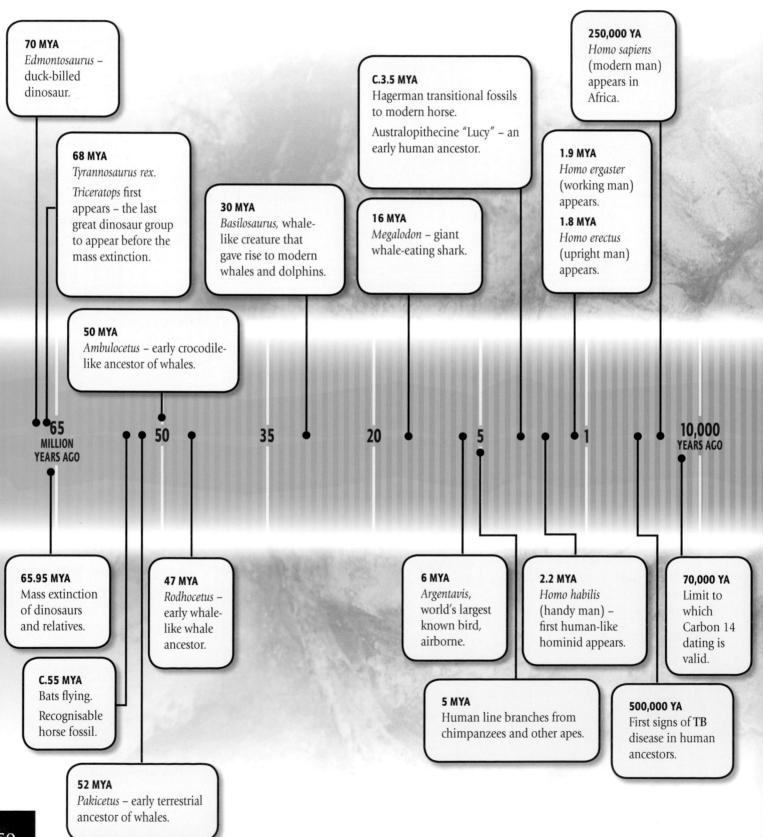

70 MYA
Edmontosaurus – duck-billed dinosaur.

68 MYA
Tyrannosaurus rex.
Triceratops first appears – the last great dinosaur group to appear before the mass extinction.

C.3.5 MYA
Hagerman transitional fossils to modern horse.
Australopithecine "Lucy" – an early human ancestor.

250,000 YA
Homo sapiens (modern man) appears in Africa.

1.9 MYA
Homo ergaster (working man) appears.

1.8 MYA
Homo erectus (upright man) appears.

30 MYA
Basilosaurus, whale-like creature that gave rise to modern whales and dolphins.

16 MYA
Megalodon – giant whale-eating shark.

50 MYA
Ambulocetus – early crocodile-like ancestor of whales.

65
MILLION
YEARS AGO

50

35

20

5

1

10,000
YEARS AGO

65.95 MYA
Mass extinction of dinosaurs and relatives.

47 MYA
Rodhocetus – early whale-like whale ancestor.

6 MYA
Argentavis, world's largest known bird, airborne.

2.2 MYA
Homo habilis (handy man) – first human-like hominid appears.

70,000 YA
Limit to which Carbon 14 dating is valid.

C.55 MYA
Bats flying.
Recognisable horse fossil.

5 MYA
Human line branches from chimpanzees and other apes.

500,000 YA
First signs of **TB** disease in human ancestors.

52 MYA
Pakicetus – early terrestrial ancestor of whales.

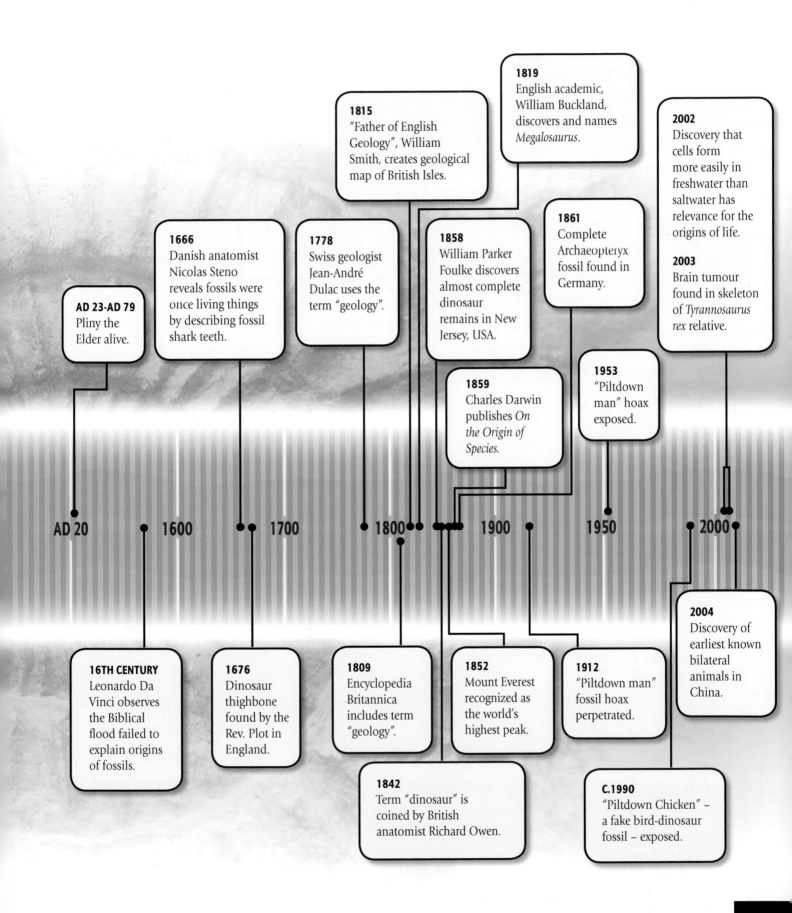

1819
English academic, William Buckland, discovers and names *Megalosaurus*.

1815
"Father of English Geology", William Smith, creates geological map of British Isles.

2002
Discovery that cells form more easily in freshwater than saltwater has relevance for the origins of life.

1666
Danish anatomist Nicolas Steno reveals fossils were once living things by describing fossil shark teeth.

1778
Swiss geologist Jean-André Dulac uses the term "geology".

1858
William Parker Foulke discovers almost complete dinosaur remains in New Jersey, USA.

1861
Complete *Archaeopteryx* fossil found in Germany.

2003
Brain tumour found in skeleton of *Tyrannosaurus rex* relative.

AD 23-AD 79
Pliny the Elder alive.

1859
Charles Darwin publishes *On the Origin of Species*.

1953
"Piltdown man" hoax exposed.

AD 20   1600   1700   1800   1900   1950   2000

16TH CENTURY
Leonardo Da Vinci observes the Biblical flood failed to explain origins of fossils.

1676
Dinosaur thighbone found by the Rev. Plot in England.

1809
Encyclopedia Britannica includes term "geology".

1852
Mount Everest recognized as the world's highest peak.

1912
"Piltdown man" fossil hoax perpetrated.

2004
Discovery of earliest known bilateral animals in China.

1842
Term "dinosaur" is coined by British anatomist Richard Owen.

C.1990
"Piltdown Chicken" – a fake bird-dinosaur fossil – exposed.

FIND OUT MORE

Further reading

National Geographic Dinosaurs (National Geographic Children's Books, 2001)

Dinosaurs!: the Biggest Baddest Strangest Fastest, Howard Zimmerman (Atheneum, 2000)

DK Guide to Dinosaurs, David Lambert (DK Children, 2006)

The Beginning of the Age of Mammals, edited by Kenneth D. Rose and David Archibald (The Johns Hopkins University Press, 2005)

Try watching

Walking with Dinosaurs (BBC Warner, 2007)

Jurassic Park Trilogy (Universal Studios, 2001)

Websites

General geology and fossils:

University of California Museum of Palaeontology

http://www.ucmp.berkeley.edu/index.php

University of California at Berkeley

http://evolution.berkeley.edu/evolibrary/home.php

London's Natural History Museum

http://www.nhm.ac.uk/

New York's American Museum of Natural History

http://www.amnh.org/

Smithsonian Institution

http://www.si.edu/

Chicago's The Field Museum

http://www.fieldmuseum.org/

Biological science news:

News of new developments in life sciences can be found at:

http://news.bbc.co.uk/2/hi/science/nature/

http://www.sciencedaily.com/

http://sciencenow.sciencemag.org/

http://www.newscientist.com/news.ns

To research

"Tongue stones"

Find out more about the history of "tongue stones". Who used them and what was it claimed they could do?

Rocks

Find examples of the different types of rocks – sedimentary, igneous, and metamorphic.

Dinosaurs

Find out more about dinosaurs. Go to your local natural history museum and draw the shapes of different species of dinosaurs.

Compare their sizes – which species were big and which were small? Which dinosaurs do you think were meat-eaters and which were plant-eaters?

What other animals were living at the same time as the dinosaurs?

"The Dinosauria" at the University of California Museum of Palaeontology's website is also a good place to start searching: http://www.ucmp.berkeley.edu/diapsids/dinosaur.html

Archaeopteryx

Explore the background to the discovery of *Archaeopteryx*. Where and how was the first specimen found? What was special about the rocks in which it was found? How were such fine details as feathers preserved?

Human evolution

Try to track the steps of human evolution from the time that we split from the other apes, about 7–8 million years ago, to the present day. Who were "Toumai", "Lucy", "Selam", the "Taung child", "Abel", "Nutcracker man", "Peking man", and "the hobbit"? When and where did they live?

"Piltdown man"

Find out more about the "Piltdown man". Where was he found and by whom? What was his impact on the scientific community at the time he was found? How was the hoax discovered? Who was involved in the hoax?

GLOSSARY

adaptive radiation   the concept of closely related species that look and behave very differently as a result of having adapted to widely differing ecological conditions

aetosaurs   heavily-armoured, crocodile-shaped, plant-eating reptiles that lived about 200 million years ago

ammonite   extinct marine molluscs, related to octopus, squid, and cuttlefish (cephalopods), which look like the living pearly nautilus

angiosperm   flowering plant

argon   a chemical element and the atmosphere's most abundant noble gas, which features in potassium-argon dating

Atlantic Conveyor   deep-water current of very salty water that travels from the Arctic Ocean to the Southern Ocean

basalt   fine-grained, dark igneous rock that flows as lava from volcanoes and cools as rock

blood corpuscles   cells in the blood

blood vessels   arteries, veins, and capillaries that carry blood

body fossil   fossils used to distinguish geological periods

Cambrian   a major division or period in the geological timescale that began about 542 million years ago and ended about 488.3 million years ago

Cambrian explosion   the rapid appearance of many different life forms during the Cambrian, which included representatives of all the phyla present on Earth today

Carboniferous   a major division or period of the geological timescale that began about 359.2 million years ago and ended about 299 million years ago

Cenozoic   an era of "new life" that covers the period between the demise of the dinosaurs 65.95 million years ago until the present day, during which all the continents moved into their current positions

co-evolve   when two organisms influence the evolution of each other

continental drift   the surface movement of Earth's major landmasses

contour feathers   large feathers that give a bird shape and colour

convergent evolution   when unrelated organisms evolve similar characteristics as adaptations to similar environmental pressures

Cretaceous   a major division or period of the geological timescale that began about 145.5 million years ago and ended about 65.95 million years ago

CT scan   "Computerized tomography scan", produced by a specialized X-ray machine which sends several beams at once into the body from different directions

cyanobacteria   bacteria that get their energy through photosynthesis

Devonian   a major division or period of the geological timescale that began about 416 million years ago and ended 359.2 million years ago

DNA   Deoxyribonucleic acid, which contains all the genetic instructions for the development and functioning of a living thing

ecological niche   a "place" in space and time that describes how an organism responds to the distribution of resources, allies, and competitors, and how it alters those factors itself

ectothermic   or "cold blooded", where an organism's heat comes from outside its body

endothermic   or "warm blooded", where an organism maintains its own heat and relatively constant body temperature regardless of the outside temperature

Eocene   a major division or epoch of the geological timescale that began about 55.8 million years ago and ended about 33.9 million years ago

eon   the largest division on the geological timescale, composed of several eras, periods, and epochs

epoch   a unit of geological timescale, less than a period but greater than an age

Equator   an imaginary line around Earth that is equidistant from the North and South Pole and divides the world into Northern and Southern hemispheres

era   a long period in the geological timescale that divides an eon into smaller units, e.g. Cenozoic Era

eurypterids   also known as sea scorpions, a group of fossil arthropods related to spiders and ticks that lived 510–248 million years ago

fibula   a calf bone in the leg attached to both ends of the tibia

filamentous   having the form of chains, threads, or filaments

fossil record   the story of life on Earth as told through fossils

fossilization   the process by which fossils are formed

granite   a medium- to coarse-textured igneous rock. Forms when molten rock "intrudes" between layers of other rocks, cools, and solidifies.

Gulf Stream   powerful, warm current in the Atlantic Ocean. It originates in the Gulf of Mexico and flows through the Florida Straits and north to Newfoundland, before crossing the Atlantic and splitting into a northerly stream that goes to northern Europe and a southerly stream that recycles back to the Americas via West Africa.

Hadean   the geological eon which started at the formation of Earth about 4,570 million years ago and ended about 3,800 million years ago

Holocene   a major division or epoch of the geological timescale that began about 11,430 years ago and continues today

igneous   rocks formed by cooled magma

Industrial Revolution   when manufacturing, transport, and agriculture changed hugely in the late 18th and early 19th centuries

isotopes   different forms of the same element

Jurassic   a major division or period of the geological timescale that started about 199.6 million years ago and ended about 145.5 million years ago

lichen   a symbiotic association of fungi and algae or cyanobacteria

lineage   group of organisms descended from a common ancestor

lobe-finned   having paired, rounded, limb-like fins

magma   molten rock that forms below Earth's surface

mantle   highly viscous layer of rock under Earth's crust

Mesozoic   a geological era that includes the "Age of the Dinosaurs"

metabolic rate   the speed with which a set of chemical reactions take place that keep an organism alive

metamorphic   pre-existing rocks that have been changed by heat and pressure

Miocene   a major division or epoch of the geological timescale that began about 23.03 million years ago and ended about 5.33 million years ago

montane forests   mountain forests

Oligocene   a major division or epoch of the geological timescale that began about 33.9 million years ago and ended about 23.03 million years ago

Ordovician   a major division or period of the geological timescale that began about 488.3 million years ago and ended about 443.7 million years ago

organism   an individual living system that reacts to stimuli, reproduces, grows, and maintains itself

ornithischia   group of beaked, plant-eating dinosaurs known as "bird-hipped dinosaurs"

ornithopods   group of bird-hipped dinosaurs that dominated the plant-eaters during the Cretaceous

Palaeocene   a major division or epoch of the geological timescale that began about 65.95 million years ago and ended about 55.8 million years ago

Palaeozoic   an era in the geological timescale spanning a period from about 542 to about 251 million years ago, to include the Cambrian, Ordovician, Silurian, Devonian, Carboniferous, and Permian periods

pelycosaurs   group of early reptiles with tall, sail-like vertebral spines

Permian   a major division or period of the geological timescale that began about 299 million years ago and ended about 251 million years ago

Phanerozoic   an overall division or eon of the geological timescale that began about 545 million years ago and continues today

phylum (plural: phyla)   a term in the classification of living things that is below "kingdom" and above "class"

phytosaurs   group of large, crocodile-shaped, predatory reptiles with long snouts that lived in the Late Triassic

"Piltdown man"   famous "early man" fossil hoax consisting of bones excavated from a gravel pit near Piltdown in East Sussex, England

plate tectonics   geological theory that explains movements of landmasses and the ocean floor

Pleistocene   a major division or epoch of the geological timescale that began about 2.59 million years ago and ended 11,430 years ago

Pliocene   a major division or epoch of the geological timescale that began about 5.33 million years ago and ended about 2.59 million years ago

potassium   chemical element important in animal nerve and brain function and plant nutrition. Used in potassium-argon dating.

Proterozoic   a geological eon extending from about 2,500 to 542 million years ago, and represented a time before complex life forms appeared

Quaternary   a major division or period of the geological timescale that extends from 2.59 million years ago to the present, and includes the Pleistocene and Holocene epochs

radioisotopic dating   technique used to date materials based on a comparison between a radioactive isotope and its decay products. Includes radiocarbon and potassium-argon dating.

rubidium   a chemical element that colours fireworks purple

saurischia   "lizard-hipped" dinosaurs, which had grasping hands and long, mobile necks. Included the giant plant-eating sauropods and the bipedal, carnivorous theropods.

sauropods   giant, long-necked and long-tailed plant-eating dinosaurs that stood on columnar legs, like elephants

sedimentary   rocks made from the break up of other rocks and their re-deposition in the sea, freshwater, or in deserts

sediments   particles of materials moved by flowing water or air

Silurian   a major division of the geological timescale that began about 443.7 million years ago and ended about 416 million years ago

sinuses   cavities (spaces) in organs or tissue. Also refers to cavities in bones of the face.

stratigraphy   study of rock layers

stromatolites   layered structures produced by cyanobacteria

strontium   a chemical element whose isotope strontium 90 is found in nuclear fallout

succession (of organisms)   predictable and orderly changes in the composition and structure of an ecological community

TB   *see* tuberculosis

Tertiary   a major division or period of the geological timescale that began with the demise of the dinosaurs 65.95 million years ago and ended at the start of the most recent major Ice Age, 2.59 million years ago. Includes the Pliocene, Miocene, Oligocene, Eocene, and Palaeocene epochs.

thecodonts   ancient reptiles. Dinosaur, bird, crocodile, and pterosaur ancestors.

theropods   bipedal, carnivorous dinosaurs

trace fossils   structures in rocks that indicate biological activity

tracheal breathing system   breathing system used by insects. An intricate series of pipes transport outside air directly to various organs in the body.

Triassic   a major division or period of the geological timescale that began about 251 million years ago and ended 199.6 million years ago

trilobite   extinct arthropod resembling a modern woodlouse.

tuberculosis (TB)   deadly infectious disease that attacks the lungs

tuffs   rocks that consist of compacted volcanic ash from an eruption

uranium   a chemical element that can be used to date rocks, as in uranium-uranium, uranium-thorium, and uranium-lead dating

X-ray   radiation with wavelengths longer than gamma rays but shorter than ultraviolet waves, used to see inside living bodies

zombie effect   fossils moved from rocks they were first deposited in and re-deposited in younger sediments. Gives a false impression that the fossils are younger.

INDEX